BASIC FUSE GUIDE

Comprehensive Guide On How To Fix All About A Fuse

Dr. Joe smith

Contents

Chapter1 3

  Introduction to anatomy of fuse 3

Chapter2 9

  Construction and working of a fuse 9

chapter3 21

  characteristics of a fuse 21

The end 31

Chapter 1

Introduction to anatomy of fuse

Anatomy of a fuse refers to the physical and functional components of a device used to protect electrical circuits from overloading and short circuits. The word "fuse" originates from the Latin word "fusus," meaning "melted," which is a key aspect of its operation. It is an essential component of any electrical system as it helps prevent damage to equipment and electrical fires. The history of fuses dates back to Thomas Edison's discovery of the electric light bulb in the 19th century. As electrical systems became widespread, the need for protection against overcurrents also increased. Initially, Edison used wires with higher resistance as fuses, but these proved to be impractical and unreliable. In 1884, John Holmes of England invented the "Holmes safety cutout," which was the first modern fuse made of a porcelain body and a lead wire. This design was subsequently improved upon by other inventors, leading to the fuses used today. The anatomy of a fuse consists of four main parts: the fuse element, the fuse body, the fuse holder, and the end caps. Let's look at these components in more detail. The Fuse Element The fuse element is the most critical component of a fuse as it is responsible for carrying the current and melting when an overcurrent occurs. It is typically made of zinc, copper, or silver, which are good conductors of electricity and have a low melting point. The diameter of the fuse element is designed to be smaller than the rest of the circuit, allowing it to heat up quickly and melt when the current exceeds its rating. The cross-sectional area and material of the fuse element determine the amount of current it can handle before melting. The Fuse Body The fuse body is the cylindrical or blade-shaped casing that houses the fuse element. It is usually made of glass, ceramic, plastic, or Bakelite, which are non-conductive materials. The purpose of the fuse body is to provide insulation and protection to the fuse element. It also has markings indicating the current rating and other relevant information about the fuse. The size and shape of the fuse body may vary depending on the type of fuse and its application. The Fuse Holder The fuse holder is the component that connects the fuse to the electrical circuit. It provides a secure and stable connection while allowing the fuse to be easily replaced when needed. The fuse holder can be a screw-type, clip-type, or a cartridge-type, depending on the type of fuse it holds. It is usually made of a conductive material such as brass, copper, or aluminum. End Caps The end caps are the metal components at the ends of the fuse body that hold the fuse element in place. They are typically made of a metal with a low melting point, such as tin, lead, or zinc. The end caps also serve as the contact points for the fuse holder. When the fuse element melts, the end caps become dislodged, and contact between the circuit and the fuse is broken, interrupting the flow of electricity.

Types of fuse

There are different types of fuses, each designed to serve a specific purpose. The most common types include the fast-acting, slow-blow, and time-delay fuses. Fast-acting fuses are commonly used in residential and commercial buildings to protect against short circuits and overloads. These fuses have a small fuse element, enabling them to react quickly to overloads. Slow-blow fuses, on the other hand, are commonly used in high-power appliances such as microwaves and refrigerators. They have thicker fuse elements, allowing them to withstand short surges of current without melting. Time-delay fuses are typically used in motors, where a high current is needed at startup. These fuses contain a spring-loaded mechanism that allows a brief surge of current before melting the fuse element

Chapter2

Construction and working of a fuse

Construction of a Fuse: The basic construction of a fuse comprises a metal strip or wire, an insulating material, and a metal casing. The metal strip or wire, known as the fuse element, is made of a low melting point metal such as copper, silver, or tin. The insulating material, usually made of ceramic, glass, or some other non-conductive material, acts as a support and prevents the fuse element from coming into contact with the metal casing. The metal casing is typically made of brass or ceramic to provide mechanical strength and to prevent damage to the fuse element. It also has a metal contact at each end, which allows the fuse to be connected to the circuit. The casing is usually cylindrical in shape and is designed in such a way that it can be easily mounted on an electrical panel or circuit board. Working of a Fuse: The working principle of a fuse is based on the heating effect of electric current. When an excessive amount of current flows through the circuit, the fuse element heats up due to its resistance to electricity. This heat causes the fuse element to melt and break, thus interrupting the current flow. This is known as blowing the fuse. The melting of the fuse element is dependent on its design and the amount of current that flows through it. Different types of fuses are designed to handle different levels of current. For instance, a fuse used in household electrical circuits is rated to handle 15-20 amps of current, while a heavy-duty industrial fuse can handle up to several hundred amps. The rating of a fuse is determined by the diameter and length of the fuse element. A longer and thinner fuse element will take longer to heat up and melt, whereas a shorter and thicker fuse element will melt quicker. This is because a longer element has more surface area and lower resistance, while a shorter element has higher resistance. The resistance of the fuse element is also affected by the material it is made of, with copper having a lower resistance than silver, and silver having a lower resistance than tin. Types of Fuses: There are various types of fuses available, each designed for specific applications based on their working voltages, current ratings, and response time. The most common types of fuses are: 1. Cartridge Fuses: These fuses are used in industrial and commercial settings. They have a cylindrical shape and are available in various sizes and current ratings. Cartridge fuses are generally more robust and can handle higher currents than other fuses. 2. Blade Fuses: These are commonly used in cars, trucks, and motorcycles. They have a plastic housing and two metal blades that are inserted into a fuse holder. The rating of a blade fuse can be easily determined by its color. 3. Glass Tube Fuses: These fuses consist of a glass tube with metal caps on each end.

They are mainly used in consumer electronic devices and come in various sizes and current ratings. 4. Resettable Fuses: Also known as polyfuse or polymer resettable fuse, these fuses are designed to reset automatically once the excessive current is removed. They have a polymer-based strip that expands with heat and breaks the connection, thus interrupting the flow of current. Once the current is back to normal, the strip cools down and reconnects the circuit. Role of Fuses in Electrical Systems: Fuses play a crucial role in ensuring the safety of electrical systems. They act as the first line of defense against overloading, short circuits, and other electrical faults. Without a fuse, an excessive amount of current can flow through the circuit, causing damage to the wiring, appliances, and even starting a fire. The fuse acts as a sacrificial component, preventing expensive and dangerous damage to the circuit and other components. Fuses are also essential for preventing electrical shocks and electrocution. In case of a fault, if a person comes into contact with the circuit, the fuse will blow, thus breaking the circuit and preventing them from getting shocked. In this way, they act as a life-saving component. Maintenance and Replacement: Fuses should be checked periodically to ensure they are in working condition. A blown fuse can be easily identified by the darkened or melted fuse element. When this happens, the fuse should be replaced with a new one of the same rating. It is important to never replace a fuse with one of a higher rating, as this can cause damage to the circuit and appliances connected to it. Regular maintenance and replacement of fuses can prevent potential hazards and ensure the safety of the electrical system.

Step 1: Understand the Purpose of Fuses
Before selecting a fuse, it is crucial to understand its purpose in a circuit. A fuse is designed to break the circuit when the current flowing through it exceeds its rated value. This prevents the excessive current from causing any damage to the system. Fuses are generally sized to protect the wiring and other components in the circuit. This is

why it is important to select the correct fuse rating to ensure proper protection of the circuit. Step 2: Determine the Type of Circuit The first step in selecting the right fuse rating is to determine the type of circuit in which the fuse will be used. There are two types of circuits - AC (alternating current) and DC (direct current). The type of circuit will affect the selection of fuse ratings as AC and DC have different characteristics. For example, in an AC circuit, the current fluctuates whereas in a DC circuit, the current remains constant. Step 3: Calculate the Maximum Current The maximum current or the full load current is the amount of current that the circuit draws when all the devices connected to it are running at their maximum capacity. This value can usually be found in the product manual or datasheet. If not, it can be calculated by adding the individual current ratings of each component in the circuit. For example, if a circuit has a light bulb that draws 2 amps and a motor that draws 3 amps, the maximum current of the circuit would be 5 amps (2 + 3 = 5). Step 4: Determine the Inrush Current Apart from the full load current, there is another type of current that should be considered when selecting the fuse rating - inrush current. This is the surge of current that is drawn when a device is first turned on. Inrush current is typically much higher than the full load current and can cause nuisance tripping of fuses. To determine the inrush current, it is recommended to consult the manufacturer's datasheet or use a specialized tool known as an inrush current tester. Step 5: Calculate the Fuse Rating Once you have determined the full load current and inrush current, the next step is to calculate the fuse rating. For AC circuits, the fuse rating should be between 125% and 200% of the full load current. This is to account for any fluctuations in the current. For DC circuits, the fuse rating should be slightly higher than the maximum current to prevent nuisance tripping. To account for the inrush current, the fuse rating should be around 150% of the inrush current. Step 6: Check the Fuse Standards It is important to note that different countries have different standards and regulations for fuses. Before selecting a fuse, it is recommended to check the standards and regulations in your country to ensure compliance. This is especially important for industrial or commercial applications where safety regulations are stringent. Step 7: Consider Ambient Temperature The temperature at which the fuse will be operating also plays a role in selecting the correct fuse rating. As the temperature increases, the resistance of the circuit decreases, which can lead to a higher current flow. If the fuse is going to be installed in an area with high temperatures, the fuse rating should be adjusted accordingly to compensate for this effect. Step 8: Consult a Professional In some cases, the application may be complex, and it may not be easy to determine the correct fuse rating. In such cases, it is recommended to consult a professional, such as an electrician or an engineer, who can help in selecting the proper fuse size for your specific application. They have the expertise and experience to make sure that the fuse selection is accurate and in compliance with safety standards.

chapter3

characteristics of a fuse

1. Current Rating: The current rating of a fuse is one of its primary characteristics. It determines the amount of current that the fuse can safely handle without tripping. Fuses are designed to operate within a specific range of current, and they are marked with their maximum current rating. If the current exceeds this rating, the fuse will melt and break the circuit, preventing further damage. 2. Interrupting Rating: Another critical characteristic of a fuse is its interrupting rating. It is the maximum fault current that the fuse can safely interrupt without causing damage. If the fault current exceeds the interrupting rating, the fuse

may catch fire, explode, or vaporize, causing significant damage to the circuit and equipment. 3. Structure and Material: Fuses are made up of a variety of materials, including glass, ceramic, plastic, and metal. The choice of material depends on the type of fuse and its application. For instance, glass fuses are used in low-voltage applications, while ceramic fuses are more durable and can withstand higher temperatures. The material used also defines the structural characteristics of the fuse and its ability to withstand thermal and mechanical stress. 4. Time-Current Curve: The time-current curve is a graphical representation of how long it takes for a fuse to trip at different current levels. Different types of fuses have different time-current curves, which are essential in selecting the right fuse for a specific application. For instance, fast-acting fuses have a steep time-current curve, meaning that they trip quickly at even small deviations from their current rating. On the other hand, slow-blow fuses have a gradual time-current curve, meaning that they can withstand currents above their rating for a short period. 5. Voltage Rating: The voltage rating is another key characteristic of a fuse. It is the maximum voltage that the fuse can withstand before breaking down and causing an electrical arc. The voltage rating is determined by the insulation and spacing between the fuse elements. Fuses with a higher voltage rating can handle higher voltages, providing better protection against short circuits and overcurrents. 6. Fuse Element: The type of material used for the fuse element also plays a crucial role in its characteristics. The most commonly used material is an alloy of tin, lead, and antimony, which has a low melting point. This alloy offers a low resistance to the electrical current, allowing it to heat up quickly and melt when the current exceeds its rating. Other materials used for fuse elements include copper, silver, and aluminum, depending on the application and specific requirements. 7. Temperature Sensitivity: The sensitivity of the fuse to temperature changes is another important characteristic. Fuses are designed to trip when exposed to excessive current, but they can also be affected by temperature changes. For instance, high temperatures caused by ambient heat or short circuits can cause the fuse element to melt prematurely, leading to false tripping. Therefore, fuses are designed to be sensitive to both electrical and thermal changes, ensuring reliable operation. 8. Indication of Failure: Fuses have a unique characteristic of showing visual indication of failure. When a fuse trips, its fuse element melts, and the failure is visible through the transparent window or indicator on the fuse body. This feature helps in quickly identifying the faulty equipment in case of an outage and replacing the fuse to restore power.

9. Resettable vs. Non-Resettable: Fuses can be resettable or non-resettable, depending on their application. Non-resettable fuses, also known as one-time fuses, need to be replaced after every trip as the fuse element melts. On the other hand, resettable fuses, also known as circuit breakers, can be reset after tripping, making them more convenient than non-resettable fuses. 10. Safety Features: There are different types of fuses available on the market, each with its unique safety features. Some fuses are designed to be tamper-proof, making them difficult to remove or replace without specialized tools. Others have built-in features such as shock hazards and fireproofing, making them safer to use in high-risk environments.

11. **Maintenance-free:** Fuses require little to no maintenance, making them a hassle-free option for protecting electrical circuits. Unlike circuit breakers, which need periodic checks and maintenance, fuses can function effectively without any regular upkeep. This reduces the overall cost and effort associated with keeping an electrical system in good working condition.

12. **Time-saving:** Fuses are time-saving devices, both in terms of installation and replacement. As mentioned earlier, fuses are easy to install, and they can also be quickly replaced in case of a fault. This allows for quick repairs and ensures minimal downtime, thus saving time and effort for both electricians and users.

13. Detect faulty circuits: Fuses have another essential characteristic - they can detect faulty circuits. When a fuse blows off, it indicates that there is an issue with the circuit, such as an overload or short circuit. This helps electricians to pinpoint and fix the problem, ensuring the safety and proper functioning of the electrical system.

14 Insulated body: Fuses have an insulated body, which prevents any contact with live or exposed electrical components. This ensures the safety of users, as well as protection against accidental shocks or electrocution. The insulation also helps to contain the heat generated during operation, preventing any potential hazards.

15. Wide operating temperature range: One of the critical characteristics of a fuse is its ability to function in a wide range of temperatures. Fuses are designed to withstand extreme temperatures, both high and low, without affecting their performance. This makes them ideal for use in different environments, such as in outdoor equipment, where temperatures can vary significantly.

16. Self-contained: Fuses are self-contained devices, meaning they do not require any external source of energy or control. They function solely based on the flow of current through the circuit, making them highly reliable and independent. This also makes fuses

suitable for use in remote locations without a power supply or in situations where a power outage may occur.

The end

www.ingramcontent.com/pod-product-compliance
Lightning Source LLC
Chambersburg PA
CBHW070923220526
45470CB00011B/1284